IMPRINT

Publisher

FMEAplus Academy GmbH
Martin Werdich
Am Engelberg 28
88239 Wangen im Allgäu
Germany
Phone +49 7522 9784 777
info@FMEAplus.de - www.FMEAplus.de

Bibliographic Information of the German National Library:

The German National Library lists this publication in the German National Bibliography; detailed bibliographic data is available on the internet at www.dnb.de.

ISBN Print: 978-3-946073-24-6 and E-Book 978-3-946073-25-3

We have created this step-by-step guide especially for self-learning courses according to the state of the art and science. We walk you step by step through all 7 phases of an FMEA project.

If you work through this workbook systematically, you will have created a complete FMEA at the end. You will then be able to create your own FMEAs.

Structure of the book:

- In each chapter, questions are first asked as a good FMEA facilitator would ask them.
- The algorithms then explain what you should do with the answers.
- The general answer shows you what this then looks like graphically
- In a simple example you can see the concrete answer graphic (The example is not complete for reasons of scope)
- In your own example, you can then practise this directly here in the workbook

This workbook is designed to get you started with very simple examples. You will realise that more complicated products require a clear presentation. Therefore, FMEA modelling is difficult to represent on paper or in tables. We recommend a suitable FMEA software for professional work.

If you want to delve deeper into this topic, we will be happy to advise, train or coach you in all aspects.

Martin Werdich and Julian Häußer wish you a lot of fun and success in your work.

Table of contents

Design FMEA 7

1 1. Definitions 7

1.1 Question collection 1 7

1.2 General answers 8

1.3 Answers example 9

1.4 Own definitions 10

2 Structural analysis 11

2.1 Question 2.1 11

2.1.1 Algorithm 11

2.1.2 General answer 11

2.1.3 Answer example 11

2.1.4 Answer own example 12

2.2 Question 2.2 13

2.2.1 Algorithm 13

2.2.2 General answer 13

2.2.3 Answer example 13

2.2.4 Answer own example 14

3 Functional analysis 15

3.1 Question3.1 15

3.1.1 Algorithm 15

3.1.2 General answer 15

3.1.3 Answer example 15

3.1.4 Answer own example 16

3.2 Question3.2 17

3.2.1 Algorithm 17

3.2.2 General answer 17

3.2.3 Answer example 17

3.2.4 Answer own example 18

3.3 Question3.3 19

3.3.1 Algorithm 19

3.3.2 General answer 19

3.3.3 Answer example 20

3.3.4 Answer own example 21
4 Error analysis 22
4.1 Question 4.1 (Focus) 22
4.1.1 Algorithm 22
4.1.2 General answer 22
4.1.3 Answer example 22
4.1.4 Answer own example 23
4.2 Question 4.2 (Consequences) 24
4.2.1 Algorithm 24
4.2.2 General answer 24
4.2.3 Answer example 25
4.2.4 Answer own example 26
4.3 Question 4.3 (Causes) 27
4.3.1 Algorithm 27
4.3.2 General answer 27
4.3.3 Answer example 28
4.3.4 Answer own example 30
4.4 Question 4.4 (completeness) 31
4.4.1 Algorithm 31
4.4.2 General answer 31
4.4.3 Answer example 32
4.4.4 Answer own example 33
5 Risk analysis 34
5.1 Question 5.1 34
5.1.1 Algorithm 34
5.1.2 General answer 34
5.1.3 Answer example 35
5.1.4 Answer own example 36
5.2 Question 5.2 37
5.2.1 Algorithm 37
5.2.2 General answer 37
5.2.3 Answer example 37
5.2.4 Answer own example 38
5.3 Question 5.3 39
5.3.1 Algorithm 39
5.3.2 General answer 39

5.3.3 Answer example ... 39
5.3.4 Answer own example ... 40
5.4 Question 5.4 ... 41
5.4.1 Algorithm ... 41
5.4.2 General answer ... 41
5.4.3 Answer example ... 42
5.4.4 Answer own example ... 43
6 Optimisation ... 44
6.1 Question 6.1 ... 44
6.1.1 Algorithm ... 44
6.1.2 General answer ... 45
6.1.3 Answer example ... 45
6.1.4 Answer own example ... 46
6.2 Question 6.2 ... 47
6.2.1 Algorithm ... 47
6.2.2 General answer ... 47
6.2.3 Answer example ... 48
6.2.4 Answer own example ... 49
7 Presentation of results ... 50
7.1 Questions 7 ... 50
7.1.1 Algorithm ... 50
7.1.2 General answer ... 51
7.1.3 Answer example ... 53
7.1.4 Answer own example ... 56
8 Appendix ... 57
8.1 B evaluation catalogue ... 57
8.2 A-rating catalogue ... 58
8.3 E-rating catalogue ... 59
8.4 AP Matrix ... 60
8.5 Form ... 61
8.6 Software ... 63

Design FMEA

FMEA is an analytical method for improving the quality, reliability and safety of products and processes. The FMEA thus supports teams in the development and manufacture of defect-free products. In addition, with a correctly conducted FMEA, one also fulfils one's duty of care as part of one's responsibility in the product development process. In this process, possible product or process defects are evaluated according to their significance, their occurrence and their detection in order to determine the need for further safeguarding measures.

The design FMEA or simply D-FMEA is often also referred to as construction FMEA. This analysis is intended to identify weak points in the constructive design of a product at an early stage and to contribute to increasing system safety, reliability and availability through optimising measures.

Your benefit is an early assessment and improvement of your product design and documents your expert knowledge.

1 1. Definitions

Objectives: The effectiveness of the entire FMEA is ensured in the definition phase and

- Helps reduce the amount of work
- is the necessary clarification of the assignment
- serves as the basis for the entire analysis.

1.1 Question collection 1

Clarification of the order:

- Aim of the FMEA?
- Type of FMEA?
- Software used (incl. version)?
- Client of the FMEA with contact address?
- Wishes of the customer?
- Prioritisation of the topics?
- Reference language?
- Which standards are relevant?
- Which evaluation catalogue?
- Scope of the FMEA (block or process flow diagram)?
- Dates, milestones?
- Core team (with rights and tasks)?
- Expanded team?
- Who invites to the meetings?
- Rules of communication and behaviour?
- Specifications?

- DVP (Design and Verification Plan)?
- Where are the meetings minuted?
- Agreements on data security (storage, archiving, data transmission)?

1.2 General answers

Clarification of the order in general:

- The aim of the FMEA is to create a general understanding of the product, risk analysis, preventive analysis and elimination of faults, documenting and securing know-how.
- Type: Design FMEA.
- Software used XXX V3.45.
- The customer of the FMEA is the buyer of the exercise book.
- The client's wish is to understand the general structure of the FMEA.
- No prioritisation is necessary here.
- The reference language is German.
- No special standards required. Method description of the AIAG/VDA2019 and the FMEAplus Academy.
- AIAG/VDA2019-Design Assessment Catalogue.
- The scope of the FMEA is the minimum required 3 levels. A block diagram is not created here.
- Error analysis Feb. 2021, risk analysis April 2021, presentation of results May 2021.
- Core team: Martin Werdich, Julian Häußer (both editing rights).
- Extended team: Stefan Dapper, Albrecht Kern, Helmut Goiginger, Mikail Bulut.
- Coordinator Sessions Martin Werdich.
- Rules of communication and behaviour according to standard method description FMEAplus Academy.
- No specifications available (not necessary here).
- No DVP (Design and Verification Plan) available (not necessary here).
- Meeting minutes are noted directly in the software.
- Agreements on data security (storage, archiving, data transmission) according to company standards.

1.3 Answers example

Clarification of the assignment Example:

- The aim of FMEA is to generate an understanding of a bell.
- Type: Design FMEA.
- Software used XXX V3.45.
- The customer of the FMEA is the buyer of the exercise book.
- The client's desire is to identify and understand the design, functions, potential defects and resulting risk of the construction of a bell using an FMEA.
- No prioritisation is necessary here as this is an example.
- The reference language is German.
- No special standards required. Method description of the AIAG/VDA2019 and the FMEAplus Academy.
- AIAG/VDA2019 Design Assessment Catalogue.
- The scope of the FMEA is the minimum required 3 levels. Block diagram is not created here.
- Error analysis Mar. 2021, risk analysis May 2021, presentation of results June 2021.
- Core team: Martin Werdich, Julian Häußer (both editing rights).
- Extended team: Stefan Dapper, Albrecht Kern, Helmut Goiginger, Mikail Bulut.
- Coordinator Sessions Martin Werdich.
- Rules of communication and behaviour according to standard method description FMEAplus Academy.
- No specifications available (not necessary here).
- No DVP (Design and Verification Plan) available (not necessary here).
- Meeting minutes are noted directly in the software.
- Agreements on data security (storage, archiving, data transmission) according to company standards.

1.4 Own definitions

2 Structural analysis

Objectives: Structural analysis is important and necessary for

- the overview of the product structure for all parties involved,
- for quick navigation during moderation, discussion of results and work with the knowledge base and as a basis for functional analysis.

2.1 Question 2.1

Which product do you want to analyse?

2.1.1 Algorithm

- Set answer as focus element
- Link to "higher-level systems
- Show Structure Tree Editor
- Mark focus element

2.1.2 General answer

Representation in the structure tree:

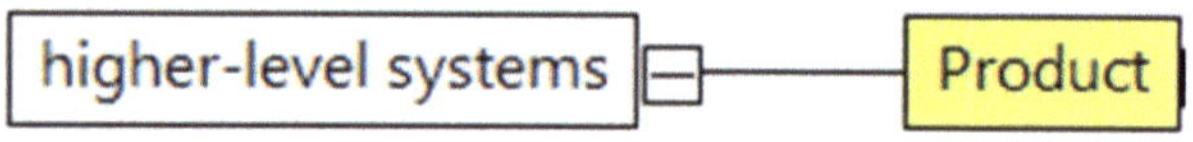

2.1.3 Answer example

Example based on a bell switch (simplified graphic)

Representation in the structure tree:

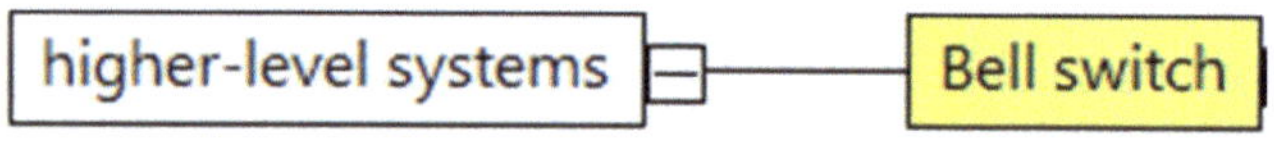

2.1.4 Answer own example

2.2 Question 2.2

Which system elements does the product consist of?

2.2.1 Algorithm

- Show the structure tree.
- Set response as subordinate system elements.

2.2.2 General answer

Representation in the hierarchical structure tree:

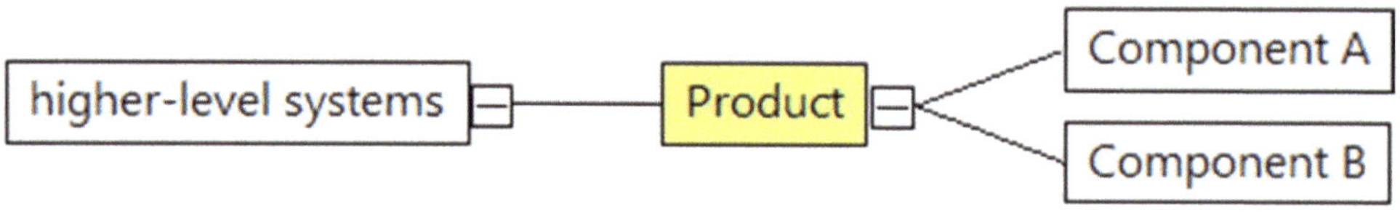

2.2.3 Answer example

Example based on a bell switch (simplified graphic)

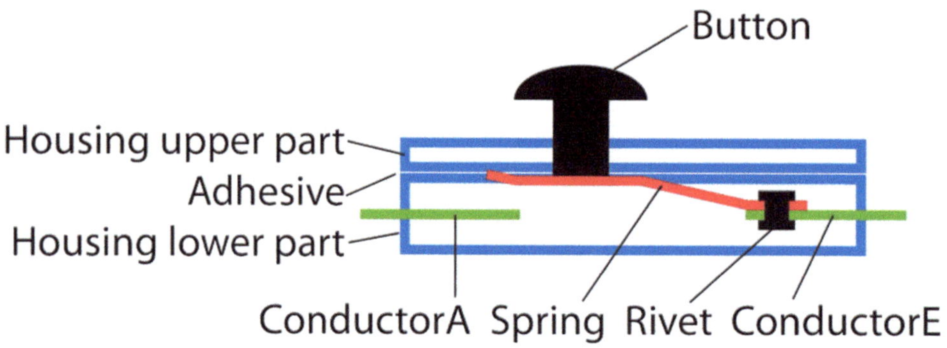

Representation in the hierarchical structure tree:

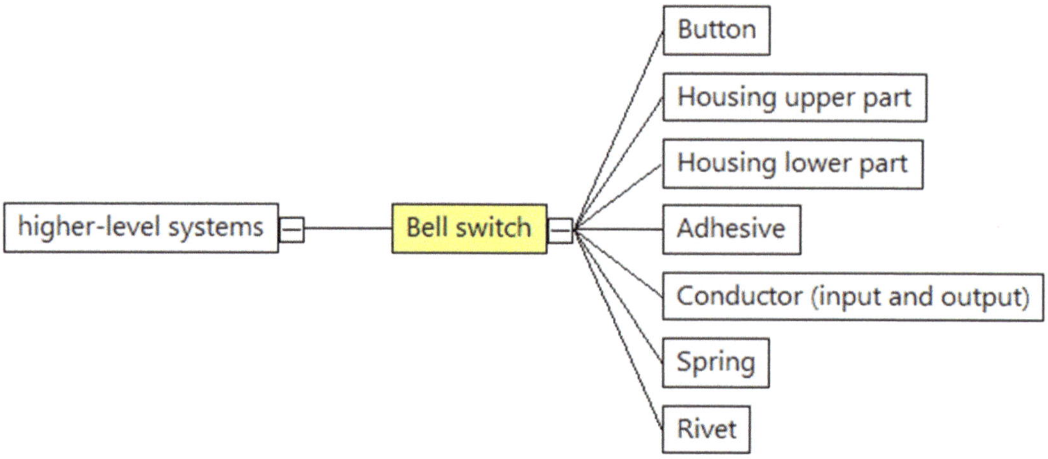

2.2.4 Answer own example

3 Functional analysis

Objectives: Functional analysis is important and necessary for

- the functional understanding of all participants and stakeholders involved in the product development process
- and as a basis for error analysis (also to minimise cause-consequence confusion).

3.1 Question3.1

What functions or/and features does your product have?

Tips:

- For the formulation of the function, you should use subject and verb.
- To help find the function, a black box method (How do I convert the inputs into the desired output?) often helps.
- Think about what the product really does, not what you want to do with it.
- The P-diagram can create understanding here.

3.1.1 Algorithm

- Hierarchically assign all functions under the focus element
- Show the first function of the focus element in the function editor

3.1.2 General answer

Hierarchical list display:

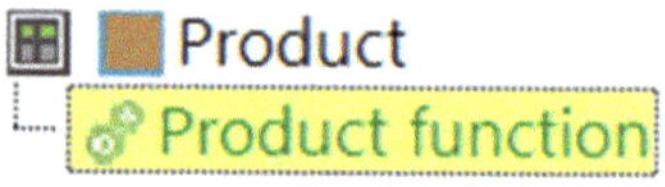

Representation in the function network:

3.1.3 Answer example

Hierarchical list display:

Representation in the function network 1:

Bell switch
Connect input and output line electrically by actuating pressure

Representation in the function network 2:

Bell switch
Without actuating pressure electrically disconnect input and output line

3.1.4 Answer own example

3.2 Question3.2

Why does the product need this function?

Tips:

- It is normal if you do not find all the higher-level functions right away. These can be added later.
- It is best to look at the FMEA of the higher-level system (if available).
- It can help if you look in the requirements or functional specifications to see why the product functions are needed.

3.2.1 Algorithm

- Enter the answer(s) hierarchically under the higher-level systems and display the link in the function editor.
- Show the next function of the focus element in the function editor
- Repeat the process until all functions have at least one sequence function

3.2.2 General answer

Hierarchical list display:

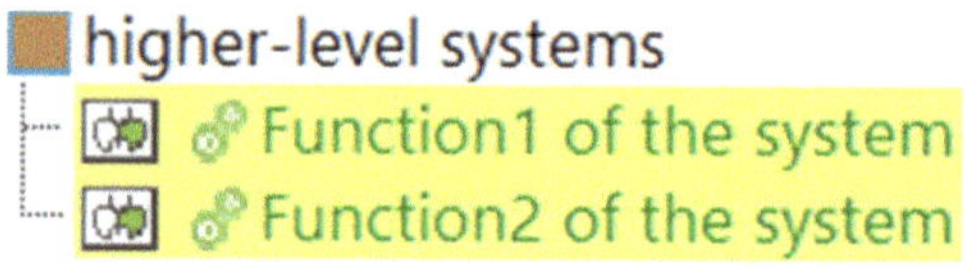

Representation in the function network:

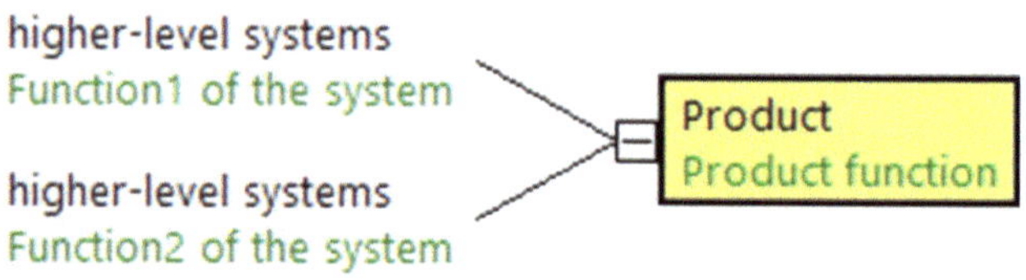

3.2.3 Answer example

Hierarchical list display:

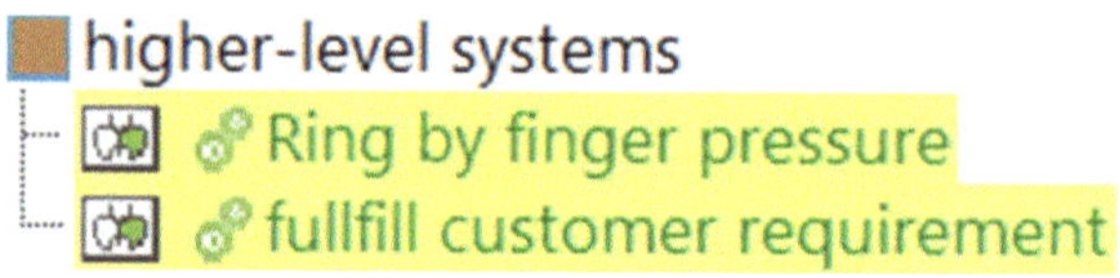

Representation in the function network 1:

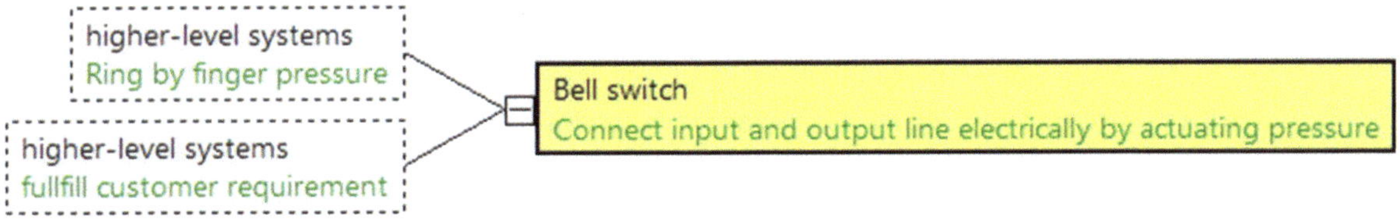

Representation in the function network 2:

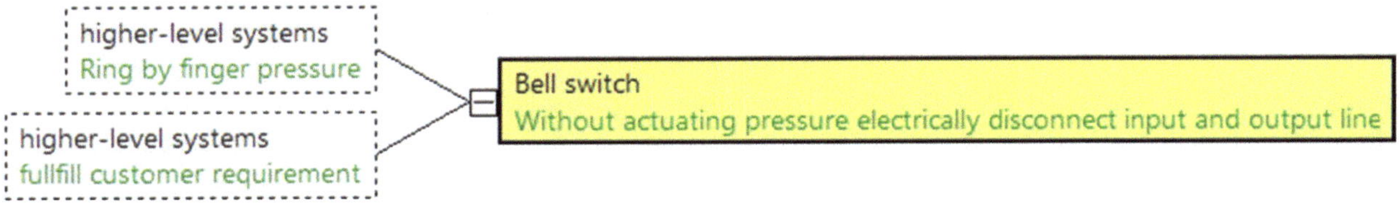

3.2.4 Answer own example

3.3 Question3.3

Which functions or product features do I need in a subordinate level to enable the functions of the focus element?

Tips:

- It is normal if you do not find all the cause functions right away. These can be added later.

3.3.1 Algorithm

- Enter the result in the subordinate system element and link it to the focus function.
- Show the overview in the function editor with the focus function in the middle.
- Repeat the process until all cause functions are linked to the focus function.
- Repeat the process until all focus functions are linked to cause functions or cause features.

3.3.2 General answer

Hierarchical list display:

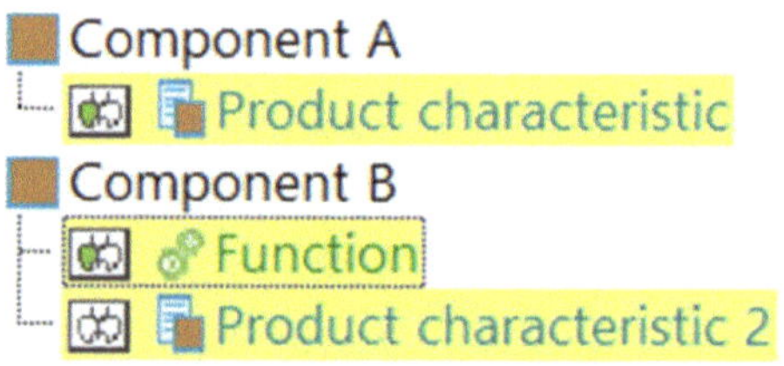

Representation in the function network:

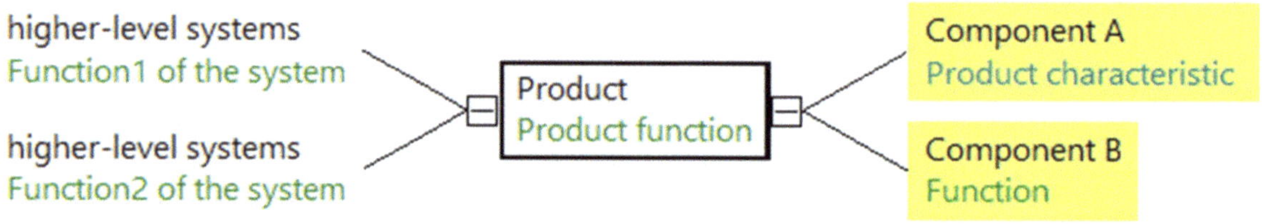

3.3.3 Answer example

Hierarchical list display:

- Button
  - Insulation resistance
  - Form
- Housing upper part
  - Insulation resistance
  - Form
- Housing lower part
  - Insulation resistance
  - Form
- Adhesive
  - Connecting the housing parts
  - Connect button with spring
- Conductor (input and output)
  - Conductivity
- Spring
  - Spring constant
  - Conductivity
- Rivet
  - Connect conductor with spring

Representation in the function network 1:

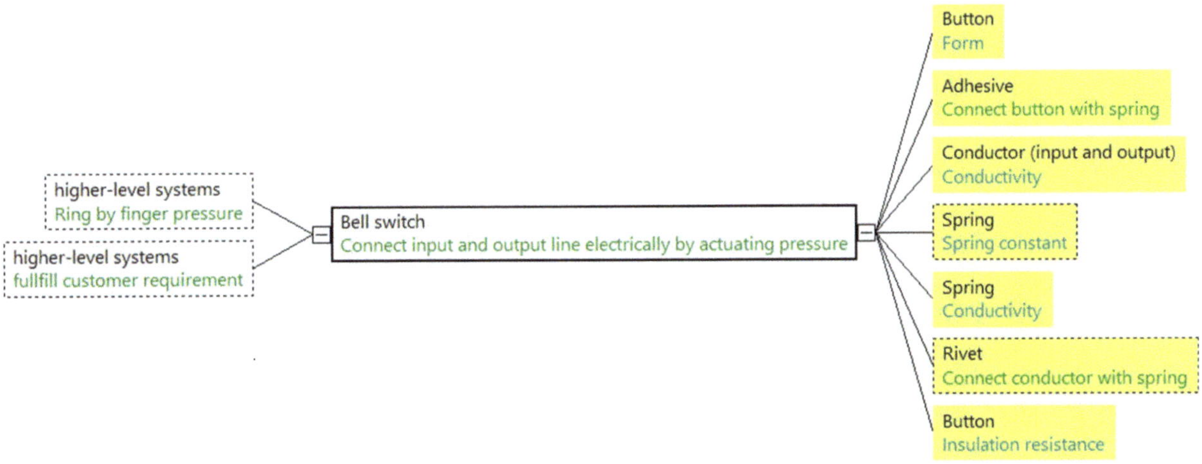

Representation in the function network 2:

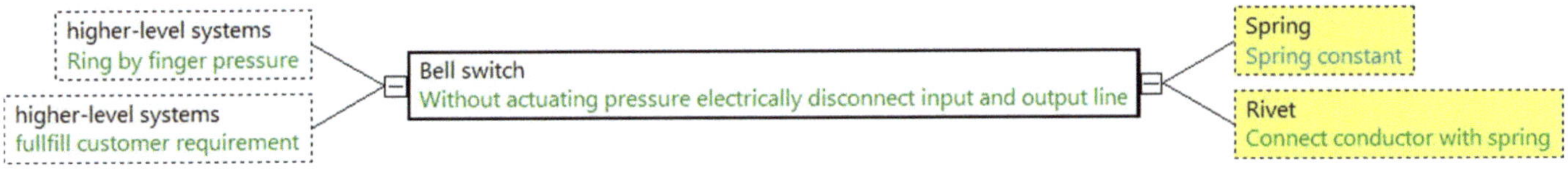

3.3.4 Answer own example

4 Error analysis

Objectives: Error analysis is important and necessary in order to

- systematically and completely find all possible errors,
- Link causes, errors and consequences
- and correctly distinguish between them.

4.1 Question 4.1 (Focus)

What are the failure modes of the focus functions?

Tips:

- Stay as close as possible to the formulation of the function.
- Systematically think through all malfunctions (no-, partial-, excessive-, intermittent- and unintended function).
- Do not note down nonsensical mistakes.

4.1.1 Algorithm

- Enter all errors hierarchically under the focus functions in the focus element.
- In the structure editor, show all errors under the focus functions in the focus element

4.1.2 General answer

Hierarchical list display of the focus level:

- Product
  - Product function
    - Complete loss of the product function
    - Product function reduced
    - excessive product function
    - temporary failure of the product function
    - unintended function

4.1.3 Answer example

Hierarchical list display of the focus level:

- Bell switch
  - Connect input and output line electrically by actuating pressure
    - Input and output line not electrically connected due to actuating pressure
    - Input and output line not completely electrically connected due to actuating pressure
    - Required actuating pressure is too weak
    - Required actuating pressure is too strong
    - Finger is electrically connected to input line
    - Bell switch falls apart
  - Without actuating pressure electrically disconnect input and output line
    - Without actuating pressure Input and output line not electrically isolated
    - Without actuating pressure Input and output line not completely electrically isolated

4.1.4 Answer own example

4.2 Question 4.2 (Consequences)

What consequences (with rating B=1-10) do the individual errors have in the higher-level systems?

Tips:

- Go through all the focus errors systematically one after the other.
- Rate the episodes individually using the B rating catalogue in the appendix.
- Think of all the levels above.
- Do not write down nonsensical consequences.

4.2.1 Algorithm

- Show the first error in the error network.
- Enter the subsequent errors with evaluation in the superordinate system under the associated function.
- Link the focus error with the associated consequence errors and show the error network.
- Enter the Bmax (highest importance of all linked sequences) as the Bmax of the error.
- After entering all the sequences, jump to the next focus error and repeat the procedures.

4.2.2 General answer

Hierarchical list representation of the sequence level:

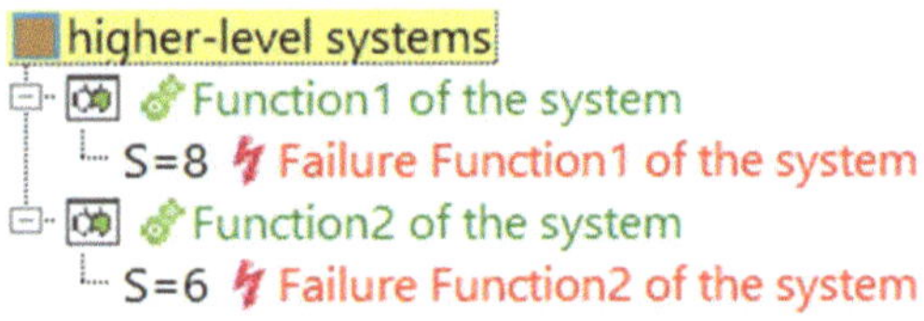

Representation in the error network:

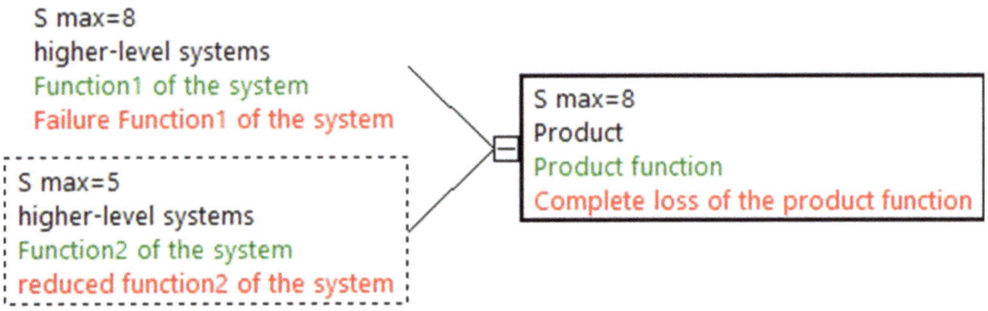

4.2.3 Answer example

Hierarchical list representation of the sequence level:

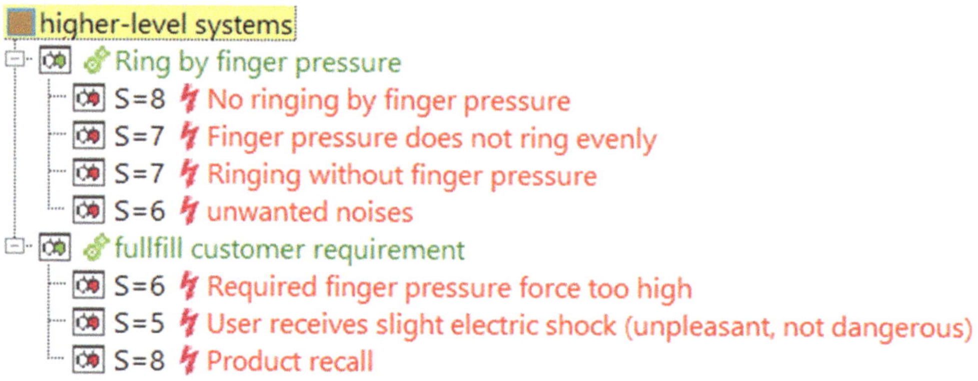

Representation in the error networks (excerpt):

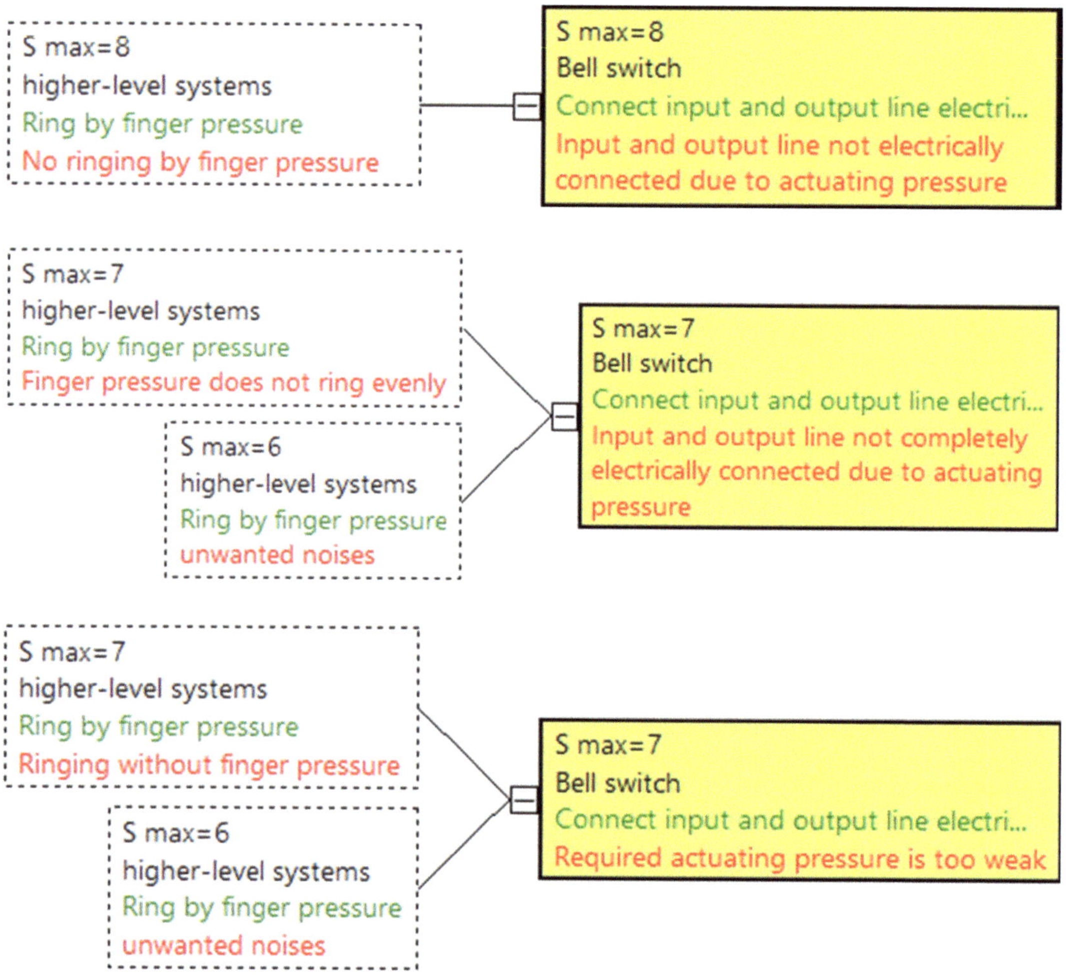

4.2.4 Answer own example

4.3 Question 4.3 (Causes)

What could be the causes of the individual errors in the subordinate system elements?

Tips:

- Go through all the focus errors systematically one after the other.
- Think of all the subordinate system elements.
- External causes are noted in a separate cause element, only if necessary. Here, the measures cannot normally be influenced, but can be noted to raise awareness of the risk and for communication purposes.
- Do not note down nonsensical causes.
- In any case, note causes that are unlikely due to design robustness.

4.3.1 Algorithm

- Show the first error in the error network.
- Enter the root cause errors (causes) in the subordinate system element under the associated function or product feature.
- Link the focus error with the associated causes and show the error network.
- Enter the Bmax (highest importance of all linked faults) as the Bmax of the cause.
- After entering all causes, jump to the next focus error and repeat the procedures.

4.3.2 General answer

Hierarchical list display of the cause levels:

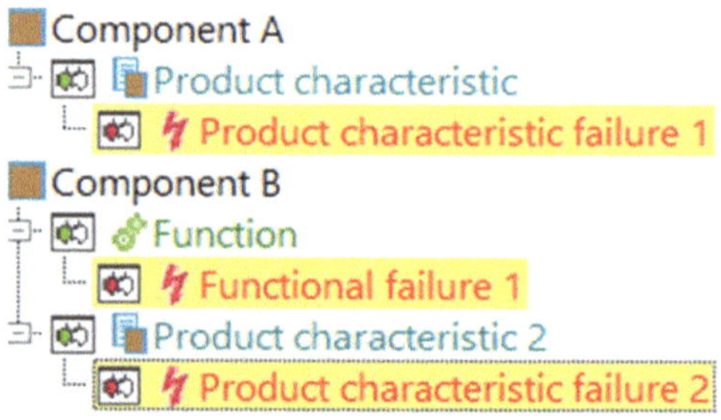

Representation in the error network 1:

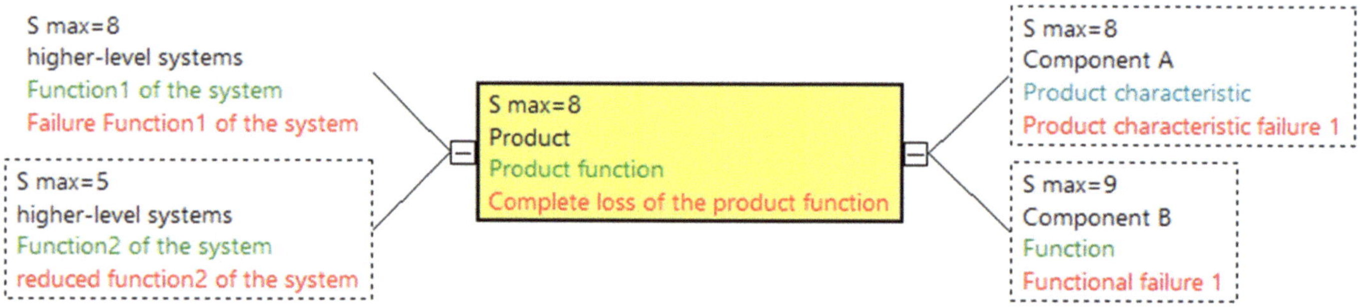

4.3.3 Answer example

Hierarchical list representation of the cause levels:

- Button
  - Insulation resistance
    - Insulation resistance too low
  - Form
    - Form unsuitable
- Housing upper part
  - Insulation resistance
    - Insulation resistance too low
  - Form
    - Form unsuitable
- Housing lower part
  - Insulation resistance
    - Insulation resistance too low
  - Form
    - Form unsuitable
- Adhesive
  - Connecting the housing parts
    - Housing parts not connected tightly enough
    - Housing parts not permanently connected
  - Connect button with spring
    - Button not connected with spring
    - Button not permanently connected to spring
- Conductor (input and output)
  - Conductivity
    - Conductivity too low
- Spring
  - Spring constant
    - Spring constant too strong
    - Spring constant too weak
  - Conductivity
    - Conductivity too low
- Rivet
  - Connect conductor with spring
    - Conductor not connected with spring
    - Conductor not permanently connected to spring

Representation in the error networks (excerpt):

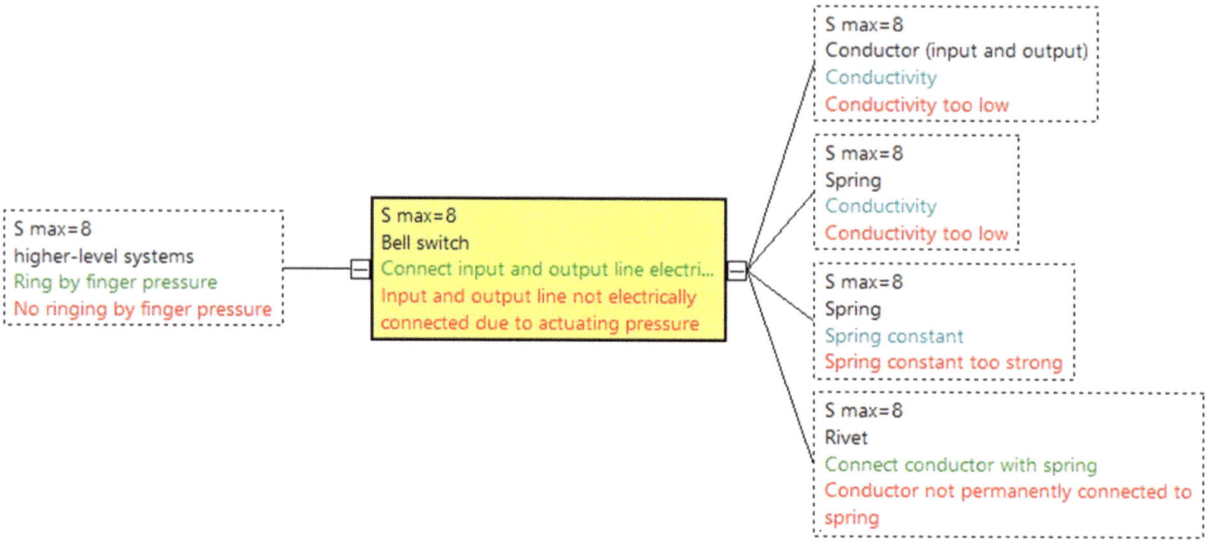

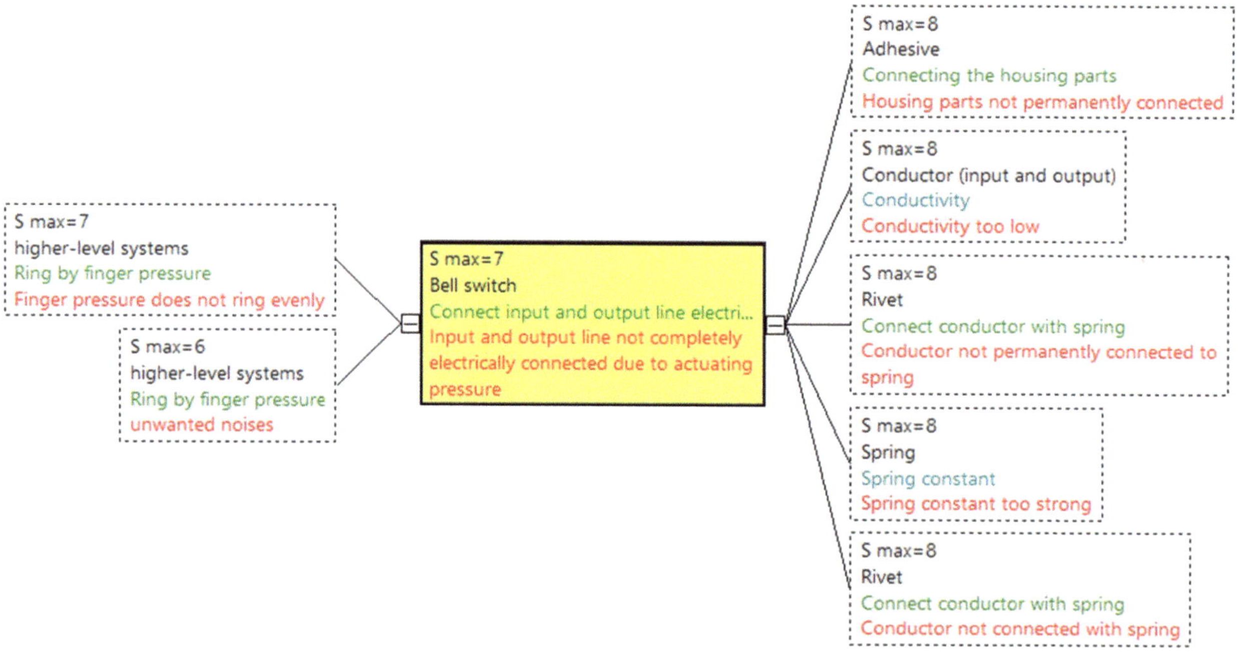

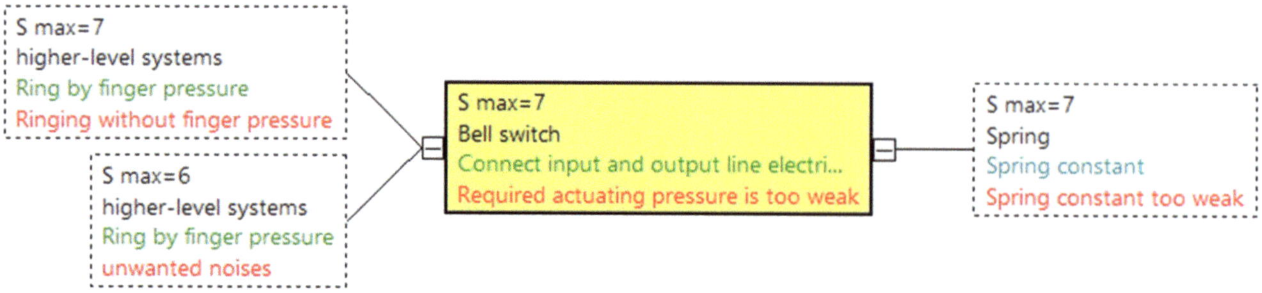

4.3.4 Answer own example

4.4 Question 4.4 (completeness)

Do you still find unlinked causes when you systematically look through all cause elements? If so, what focus error could be a possible consequence?

Tips:

- With this step you make sure that you have not forgotten anything.
- You can also brainstorm with the component experts about other possible causes.

4.4.1 Algorithm

- Show first cause system element.
- Click on the link icon of the cause to show the error network with cause focus.
- Link the cause to the corresponding error in the focus level.
- If requested, show the complete error network for verification in the team.

4.4.2 General answer

Hierarchical representation of the cause element:

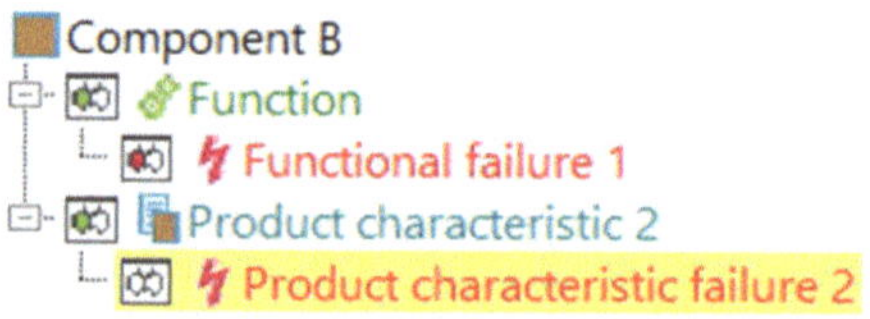

Error network with focus on cause:

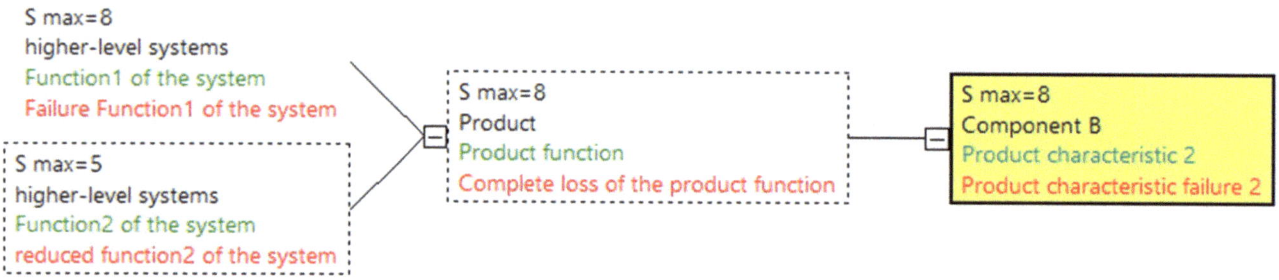

In the next step, you will see the complete error network with focus on the error:

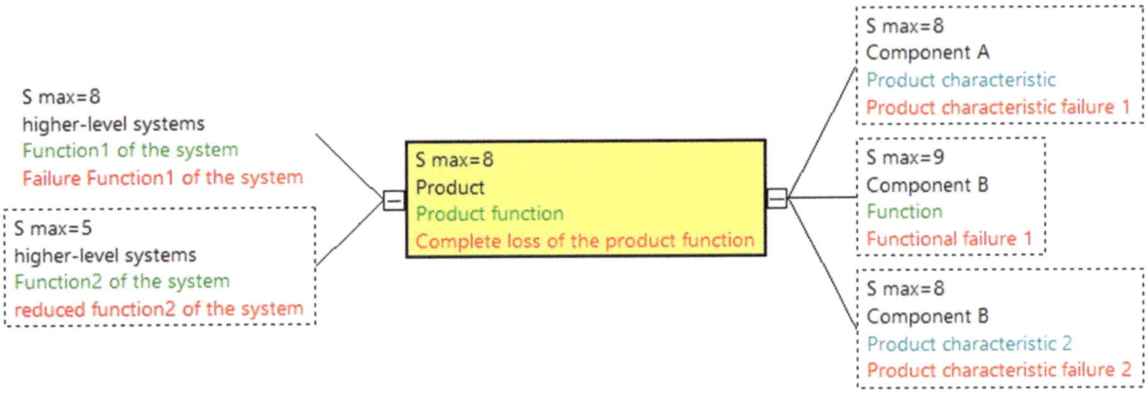

4.4.3 Answer example

Hierarchical representation of the first cause element:

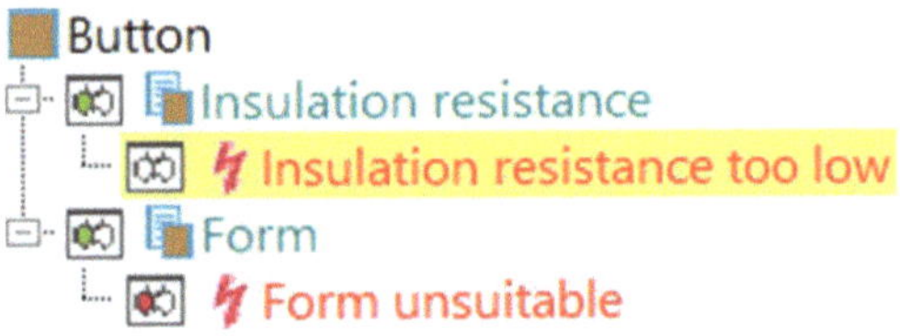

Error network with focus on cause:

(in the first case, a defect had to be added to the focus. This happens again and again, as the completeness of the nets constantly increases through the systematic processing in the course of the FMEA).

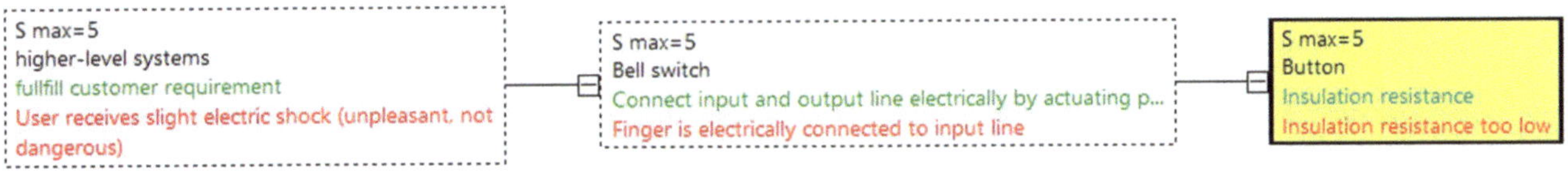

In the next step, after processing all causes, you will see the complete error network with focus on the error (this is then used for verification by the team):

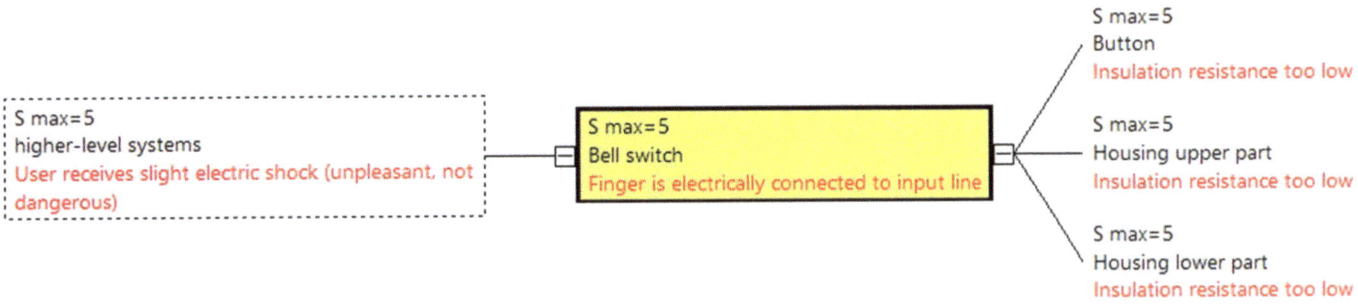

4.4.4 Answer own example

5 Risk analysis

Objectives: Risk analysis is important and necessary to

- systematically determine the actual risk situation
- and to quickly determine the need for optimisation in a comprehensible way.

5.1 Question 5.1

What have you done constructively or what will you do by default to make the cause less likely to occur?

Tips:

- Think about all the measures you take before creating the prototype. E.g. calculations, simulations, adoptions, benchmark, design according to specification design manual, ...).

5.1.1 Algorithm

- Show the cause-based error network for orientation.
- Show the first cause of error in the list view and the cause-based error network.
- Insert the "initial measure status" here under the cause.
- Enter the response as an avoidance measure below the initial measure level here in the list view.

5.1.2 General answer

Cause-based error network (for orientation and info for the team):

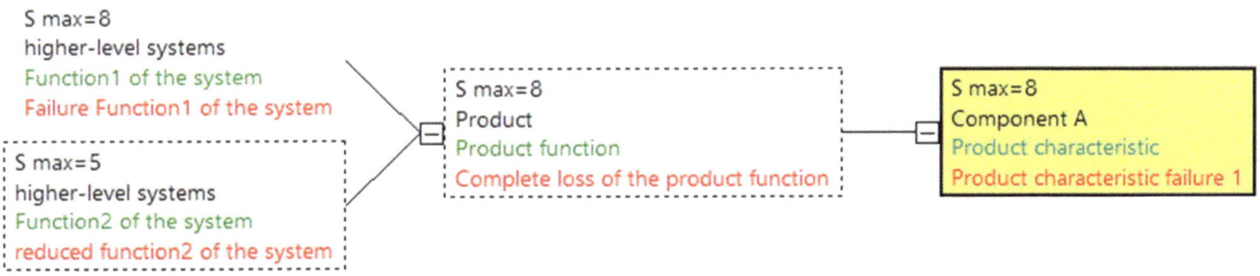

List view:

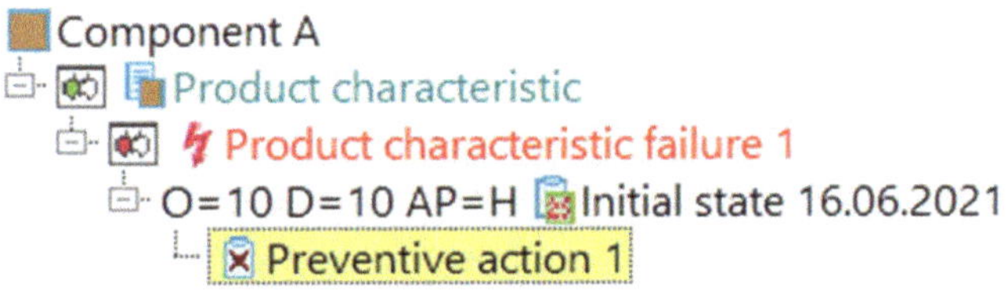

5.1.3 Answer example

Cause-based error network (for orientation and info for the team):

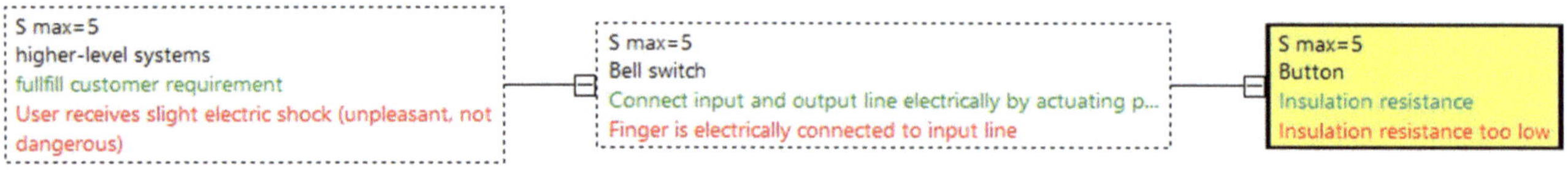

List view:

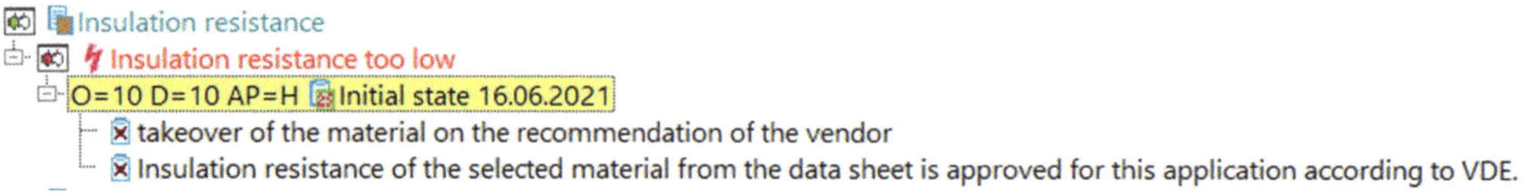

5.1.4 Answer own example

5.2 Question 5.2

How do you currently rate the A (occurrence) of the cause if you include all your prevention measures?

Tips:

- Rate using the A-rating catalogue in the appendix.

5.2.1 Algorithm

- Enter the A rating as an attribute in the initial measure state.
- Show this in the list view.

5.2.2 General answer

List view:

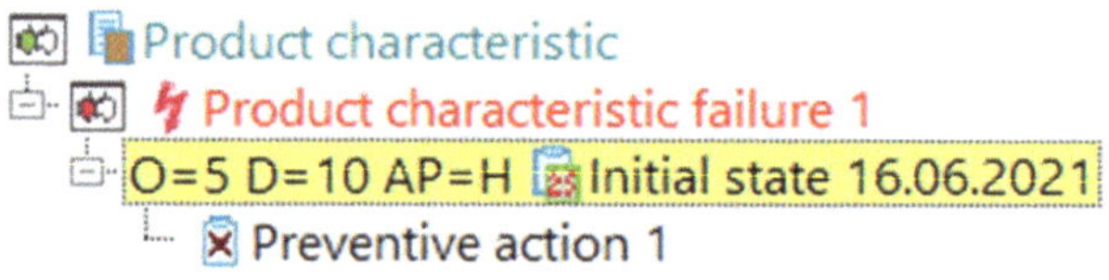

5.2.3 Answer example

List view:

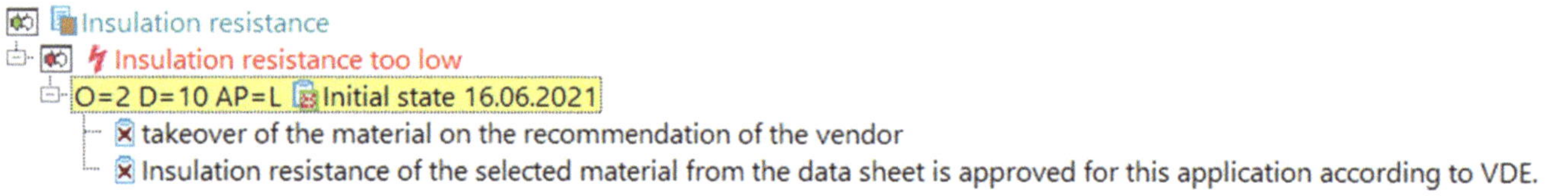

5.2.4 Answer own example

5.3 Question 5.3

If the cause were to occur, what detection measure would enable you to identify it or its consequences in time?

Tips:

- Think of all the discovery activities (verifications and validations) that you perform with the prototype. E.g. trial assemblies, load tests, endurance runs, system tests, ...).

5.3.1 Algorithm

- Enter the response as a discovery measure below the initial measure level in hierarchical form under the list view.

5.3.2 General answer

List view:

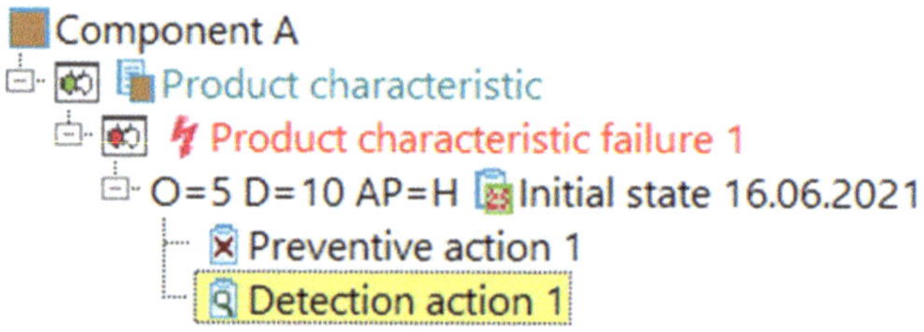

5.3.3 Answer example

List view:

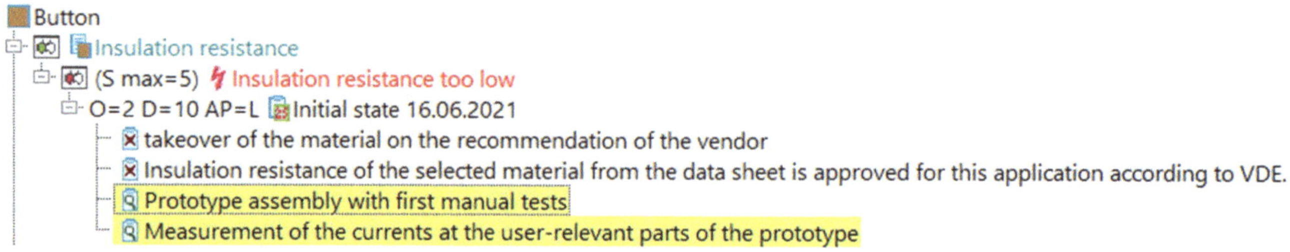

5.3.4 Answer own example

5.4 Question 5.4

How do you currently rate the E (discovery) of the cause or its consequences when you include all your discovery measures?

Tips:

- Rate using the E rating catalogue in the appendix.

5.4.1 Algorithm

- Enter the E rating as an attribute in the initial measure state.
- Enter the resulting AP (Action Priority) as an attribute in the initial measure status.
- Synchronise the form regularly (error-centred and at all levels with context to the associated function and system element):
- Error in the error column
- Linked episodes in the episode column
- Linked causes in the cause column
- Avoidance measures and detection measures, as well as their evaluation in their columns under the initial measure status of the cause.
- Show the result in the list view and in the form.
- After entering all measures, jump to the next cause and repeat the risk analysis procedures.

5.4.2 General answer

List view:

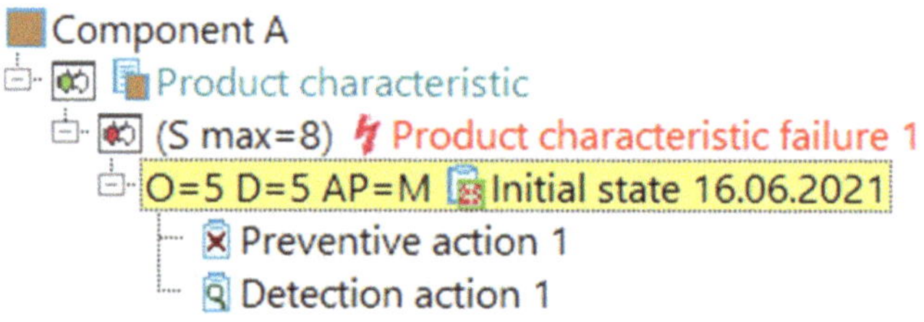

Form view for control:

Effect	S	Failure mode	Cause	Preventive action	O	Detection action	D	AP
System element: Product								
Function: [Product] **Product function**								
[higher-level systems] <Function1 of the system> >> Failure Function1 of the system	8	Complete loss of the product function	[Component A] <Product characteristic > >> Product characteristic failure 1	Initial state: 16.06.2021				
				Preventive action 1	5	Detection action 1	5	M
[higher-level systems] <Function2 of the system> >> reduced function2 of the system	5		[Component B] <Function> >> Functional failure 1	Initial state: 16.06.2021				
				Preventive action2	3	Detection action 2	3	L
			[Component B] <Product characteristic 2> >> Product characteristic failure 2	Initial state: 16.06.2021				
				Preventive action2	2	Detection action 2	2	L

5.4.3 Answer example

List view:

- Insulation resistance
  - (S max=5) Insulation resistance too low
    - O=2 D=2 AP=L Initial state 16.06.2021
      - takeover of the material on the recommendation of the vendor
      - Insulation resistance of the selected material from the data sheet is approved for this application according to VDE.
      - Prototype assembly with first manual tests
      - Measurement of the currents at the user-relevant parts of the prototype

Form view for control:

System element: Bell switch								
Function: [Bell switch] **Connect input and output line electrically by actuating pressure**								
[higher-level systems] <fullfill customer requirement> >> User receives slight electric shock (unpleasant, not dangerous)	5	Finger is electrically connected to input line	[Button] <Insulation resistance> >> Insulation resistance too low	Initial state: 16.06.2021 takeover of the material on the recommendation of the vendor Insulation resistance of the selected material from the data sheet is approved for this application according to VDE.	2	Prototype assembly with first manual tests Measurement of the currents at the user-relevant parts of the prototype	2	L

5.4.4 Answer own example

6 Optimisation

Objectives: Optimisation is important and necessary to

- Identify the risk at the end of the PEP (product development process) in the best possible way,
- propose and pursue additional measures in a timely manner,
- and systematically reduce the risk for all parties involved.

6.1 Question 6.1

By which additional measures (prevention and/or detection) would it be possible to further reduce the risk of the cause?

Tips:

- If AP=H (High), additional measures **must be** taken or a justification noted.
- If AP=M (Middle), additional measures **should be** taken or a justification noted.
- If AP=L (Low), additional measures **can be** taken.

- Take into account the state of the art and science, as well as the costs.
- Use comments so that your decisions can be understood.
- Add a responsible person, a deadline and a status to each optimising measure so that it can be tracked by project management.
- One FMEA team member should be responsible.
- The date should be chosen in such a way that it can be realistically implemented and the measure can still be carried out in time.
- Statuses that make sense are:
- unedited
- in decision
- In implementation
- completed
- discarded

6.1.1 Algorithm

- Show the causes with the AP=H and AP=M
- Show the first cause of error to be optimised in the list view.
- Insert the "optimising measure level" here under the cause and after the initial measure level.

- Enter the optimising avoidance measure as an avoidance measure below the optimising measure status hierarchically in the list view and add V, T, S (responsible person, deadline, status) as an attribute to the measure.
- Enter the optimising detection measure as a detection measure below the optimising measure status hierarchically in the list view and add V, T, S (responsible person, deadline, status) as an attribute to the measure.

6.1.2 General answer

List view:

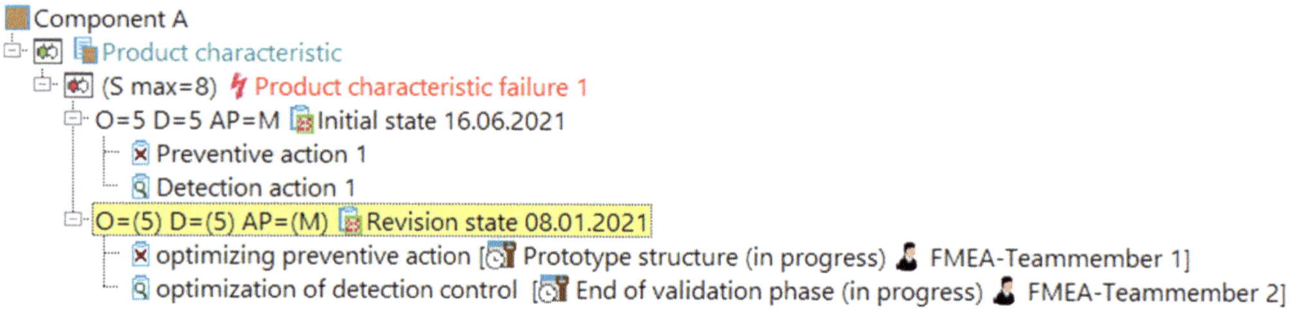

6.1.3 Answer example

List view:

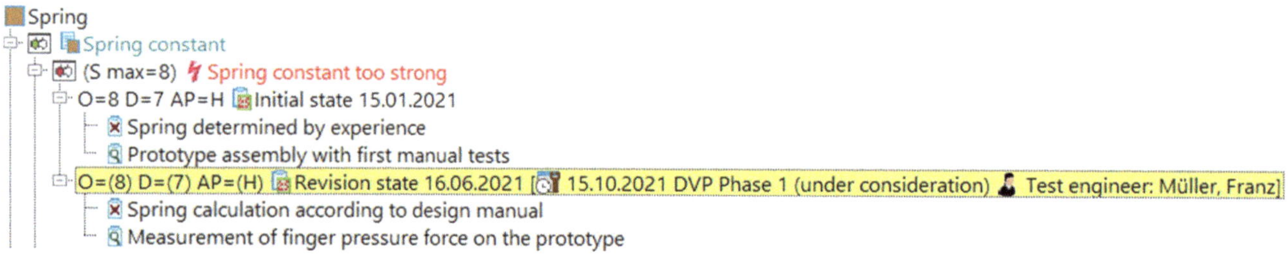

6.1.4 Answer own example

6.2 Question 6.2

Taking these measures into account, how would you rate the cause in the assessments (future) if they were implemented?

Tips:

- Evaluate as a team. in order to safeguard yourself and to get values that are as objective as possible.
- After successful implementation of the optimising measures, note the verified and validated result in the FMEA and adjust the ratings A and E if necessary.

6.2.1 Algorithm

- Enter the A and E rating as an attribute in the optimising measure state.
- Show this in the list view and the form.
- After entering the optimised score, jump to the next cause to be optimised and repeat the optimisation procedures.

6.2.2 General answer

List view:

Component A
Product characteristic
(S max=8) Product characteristic failure 1
O=5 D=5 AP=M Initial state 16.06.2021
Preventive action 1
Detection action 1
O=(3) D=(2) AP=(L) Revision state 08.01.2021
optimizing preventive action [Prototype structure (in progress) FMEA-Teammember 1]
optimization of detection control [End of validation phase (in progress) FMEA-Teammember 2]

Form view for control:

Effect	S	Failure mode	Cause	Preventive action	O	Detection action	D	AP
System element: Product								
Function: [Product] **Product function**								
[higher-level systems] <Function1 of the system> >> Failure Function1 of the system	8	Complete loss of the product function	[Component A] <Product characteristic > >> Product characteristic failure 1	Initial state: 16.06.2021				
				Preventive action 1	5	Detection action 1	5	M
				Revision state: 08.01.2021				
[higher-level systems] <Function2 of the system> >> reduced function2 of the system	5			optimizing preventive action FMEA-Teammember 1 Prototype structure (in progress)	(3)	optimization of detection control FMEA-Teammember 2 End of validation phase (in progress)	(2)	(L)

6.2.3 Answer example

List view:

- Spring
  - Spring constant
    - (S max=8) Spring constant too strong
      - O=8 D=7 AP=H Initial state 15.01.2021
        - Spring determined by experience
        - Prototype assembly with first manual tests
      - O=(3) D=(4) AP=(L) Revision state 16.06.2021 [15.10.2021 DVP Phase 1 (under consideration) Test engineer: Müller, Franz]
        - Spring calculation according to design manual
        - Measurement of finger pressure force on the prototype

Form view for control:

Effect	S	Failure mode	Cause	Preventive action	O	Detection action	D	AP	R/D
System element: Bell switch									
Function: [Bell switch] **Connect input and output line electrically by actuating pressure**									
[higher-level systems] <Ring by finger pressure> >> No ringing by finger pressure	8	Input and output line not electrically connected due to actuating pressure	[Spring] <Spring constant> >> Spring constant too strong	Initial state: 15.01.2021					
				Spring determined by experience	8	Prototype assembly with first manual tests	7	H	
				Revision state: 16.06.2021					
				Spring calculation according to design manual	(3)	Measurement of finger pressure force on the prototype	(4)	(L)	Test engineer: Müller, Franz 15.10.2021 DVP Phase 1 under consideration

6.2.4 Answer own example

7 Presentation of results

Objectives: The benefit of the entire FMEA is recognised in the presentation of results. This is created as a clear notation, reduced to the most necessary, individually for the target group and

- serves as an overview of project risks for decision-makers, clients, project participants and interested parties,
- is the basis of their decisions,
- as well as an applicable document in the product development process
- and is versioned and archived together with the entire PEP documentation.

7.1 Questions 7

- Who is the presentation for?
- What is important for the recipient?
- What results does the recipient need for his work?

Tips:

- It is best to create a summary as a one-pager before the cover page.
- Caution with secret or sensitive data (possibly clarify with higher-level decision-makers beforehand).

7.1.1 Algorithm

- Create an overview of the results using frequency analysis AP.
- List the causes with AP=H and AP=M.
- List the measures with assessment as context to the causes.
- Store versioned documentation in a secure and retrievable manner. This should contain the following:
- the overviews
- the cover sheet
- the FMEA form (preferably small VDA A4 portrait)
- the FMEA software file
- All related and necessary documents (protocols, drawings, BM list, block diagram, test results, ...)
- possibly links to the project content.

7.1.2 General answer

A complete general documentation would go beyond the scope of this book. Therefore, some exemplary graphics follow as suggestions.

The structural analysis:

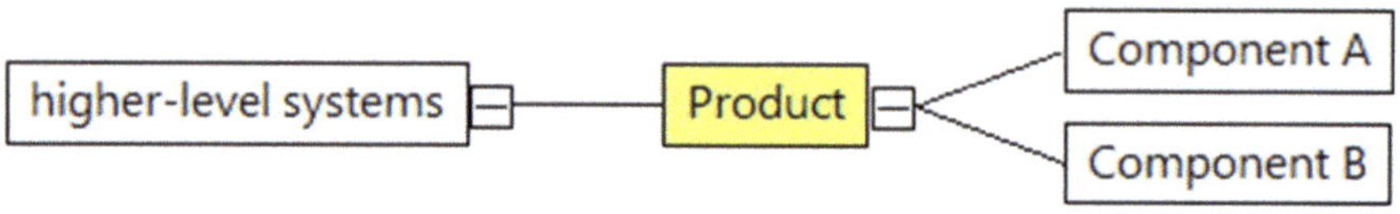

Possible errors of the product with the maximum significance:

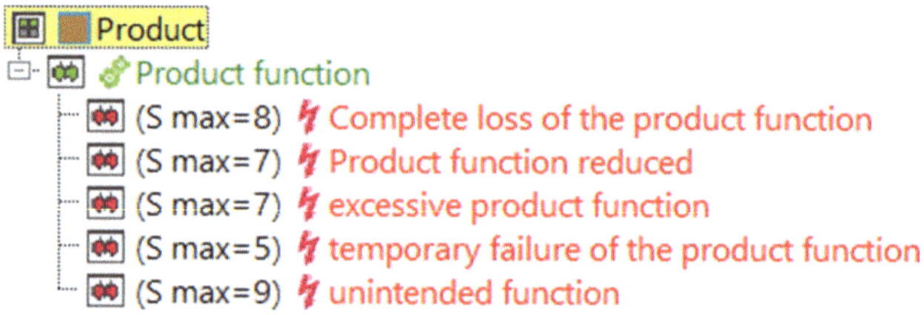

Possible consequences of the product with evaluation:

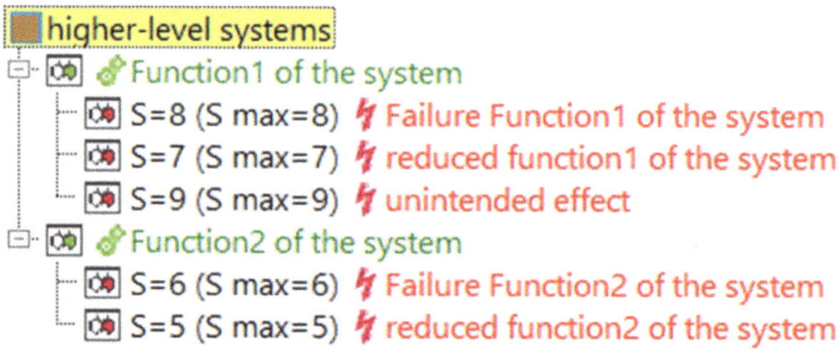

Frequency analysis AP (blue = initial measure status, green = optimisation status:

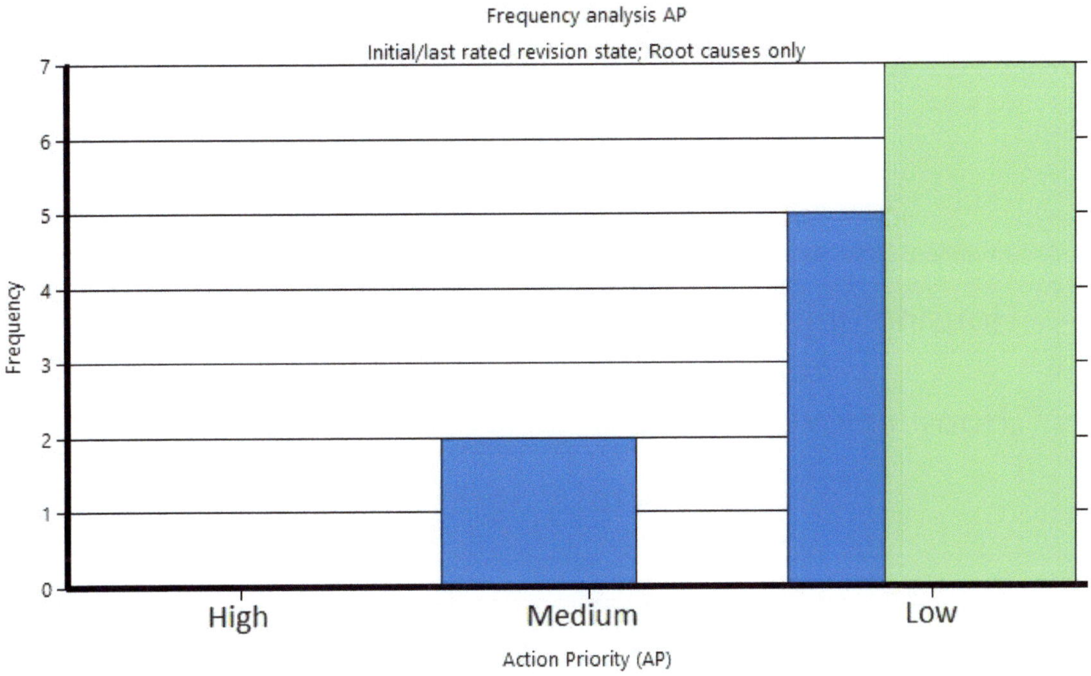

Causes with AP=H and M and their actions in a cause-centred form:

FMEA plus AKADEMIE			Causes based FMEA Form Component A					
Effect	S	C	Failure mode	Cause	Preventive action	O	Detection action	D
[higher-level systems] <Function1 of the system> Failure Function1 of the system	8		[Product] <Product function> Complete loss of the product function	[Component A] <Product characteristic > Product characteristic failure 1	Initial state: 16.06.2021 Preventive action 1	5	Detection action 1	5
[higher-level systems] <Function2 of the system> reduced function2 of the system	5		[Product] <Product function> Product function reduced		Revision state: 08.01.2021 optimizing preventive action V: FMEA-Teammember 1 Prototype structure (in progress)	(3)	optimization of detection control V: FMEA-Teammember 2 End of validation phase (in progress)	(2)
[higher-level systems] <Function1 of the system> reduced function1 of the system	7		[Product] <Product function> temporary failure of the product function					

7.1.3 Answer example

A complete example documentation would go beyond the scope of this book. Therefore, some exemplary graphics follow as suggestions.

Product image:

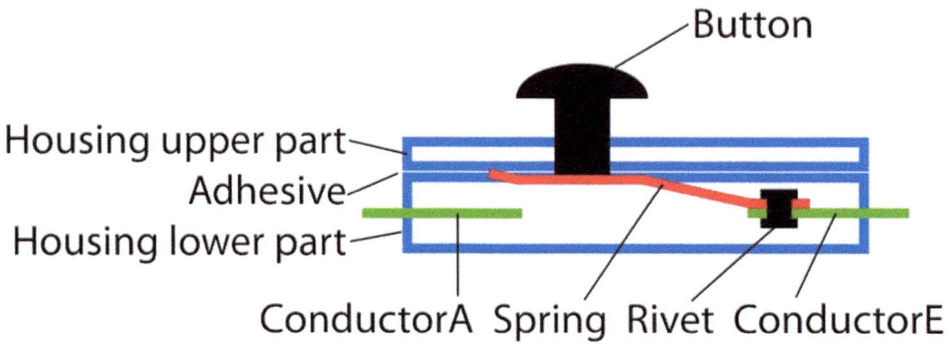

The structural analysis:

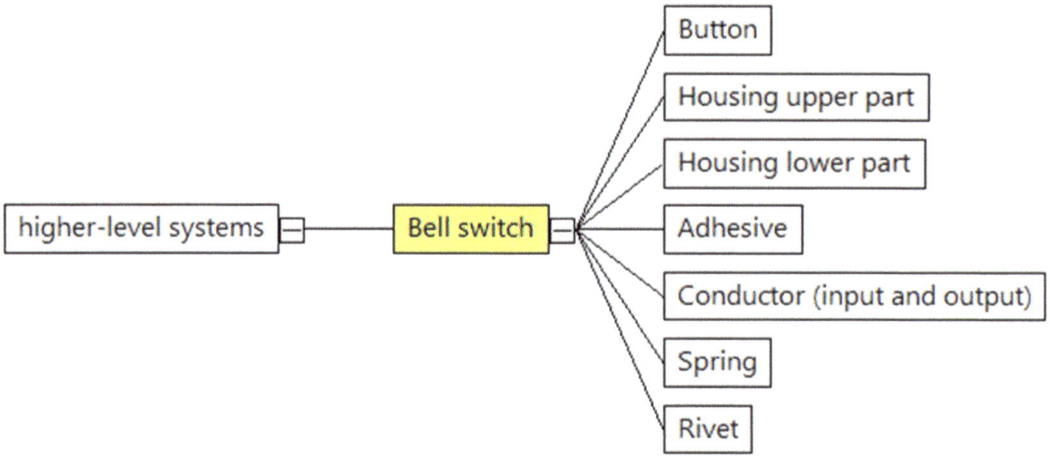

Possible errors of the product with the maximum significance:

- Bell switch
  - Connect input and output line electrically by actuating pressure
    - (S max=8) Input and output line not electrically connected due to actuating pressure
    - (S max=5) Finger is electrically connected to input line
    - (S max=7) Input and output line not completely electrically connected due to actuating pressure
    - (S max=7) Required actuating pressure is too weak
    - (S max=8) Required actuating pressure is too strong
    - (S max=8) Bell switch falls apart
  - Without actuating pressure electrically disconnect input and output line
    - (S max=7) Without actuating pressure Input and output line not electrically isolated
    - (S max=7) Without actuating pressure Input and output line not completely electrically isolated

Possible consequences of the product with evaluation:

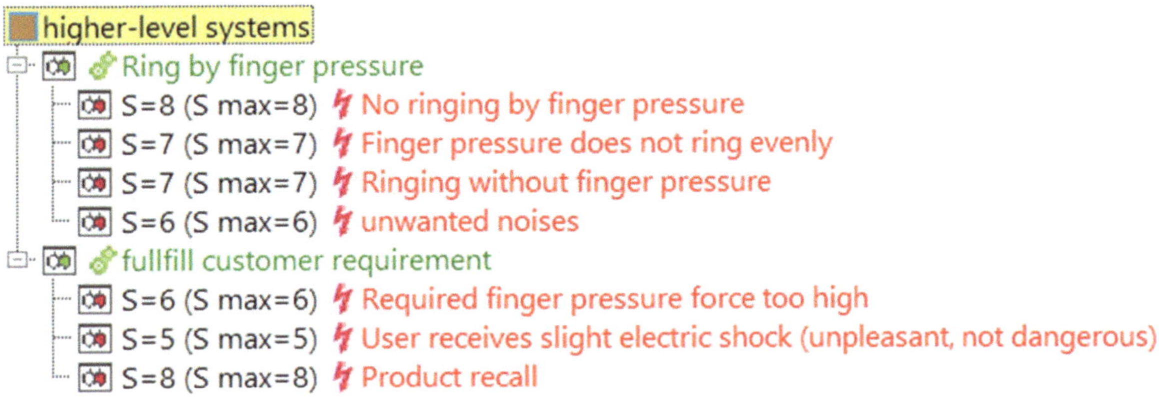

Frequency analysis AP (blue = initial measure status, green = optimisation status:

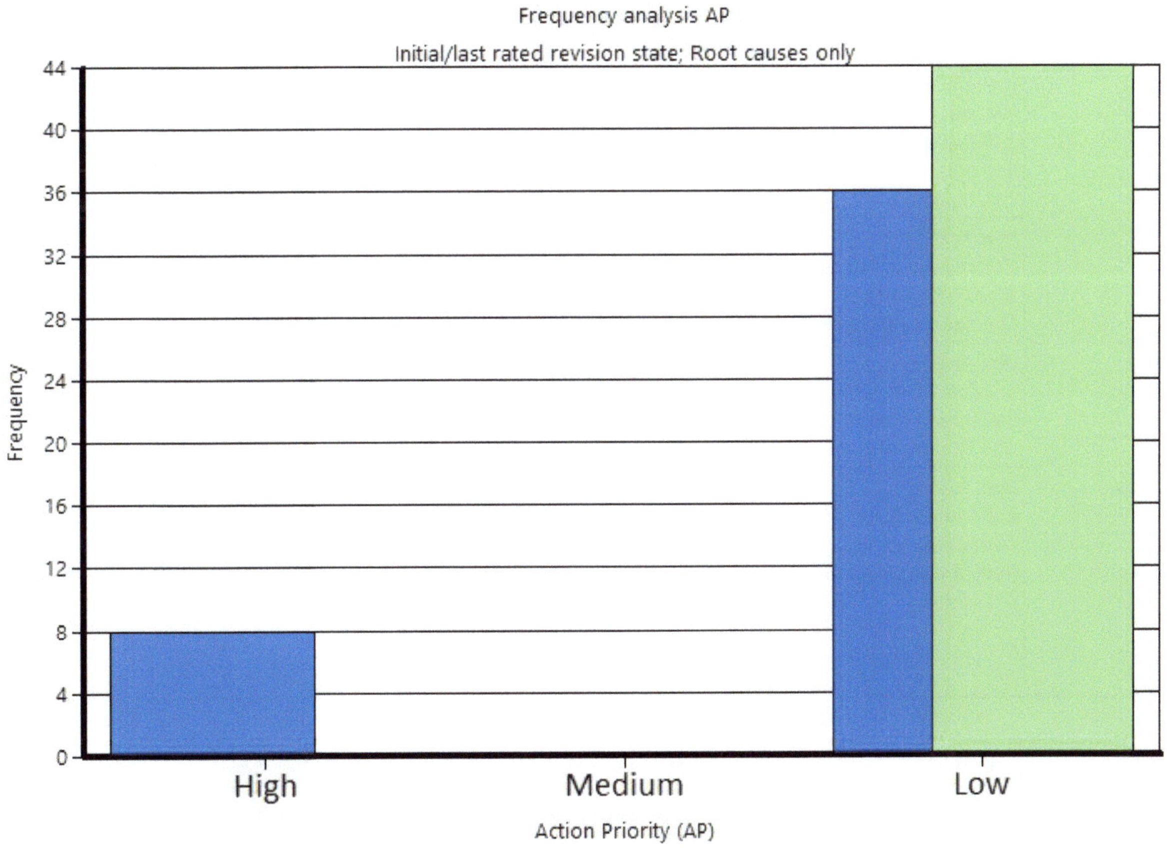

Causes with AP=H and M and their actions in a cause-centred form:

FMEA plus AKADEMIE	Causes based FMEA Form Spring						
Effect	S	Failure mode	Cause	Preventive action	O	Detection action	D
[higher-level systems] No ringing by finger pressure	8	[Bell switch] Input and output line not electrically connected due to actuating pressure	[Spring] Spring constant too strong	Initial state: 15.01.2021 Spring determined by experience	8	Prototype assembly with first manual tests	7
[higher-level systems] Finger pressure does not ring evenly	7	[Bell switch] Input and output line not completely electrically connected due to actuating pressure		Revision state: 16.06.2021 Spring calculation according to design manual	(3)	Measurement of finger pressure force on the prototype	(4)
[higher-level systems] unwanted noises	6						
[higher-level systems] Required finger pressure force too high	6	[Bell switch] Required actuating pressure is too strong					
[higher-level systems] Ringing without finger pressure	7	[Bell switch] Required actuating pressure is too weak	[Spring] Spring constant too weak	Initial state: 15.01.2021 Spring determined by experience	8	Prototype assembly with first manual tests	7
[higher-level systems] unwanted noises	6	[Bell switch] Without actuating pressure Input and output line not completely electrically isolated		Revision state: 15.01.2021 Spring calculation according to design manual	(3)	Measurement of finger pressure force on the prototype	(4)
		[Bell switch] Without actuating pressure Input and output line not electrically isolated					
[higher-level systems] No ringing by finger pressure	8	[Bell switch] Input and output line not electrically connected due to actuating pressure	[Spring] Conductivity too low	Initial state: 15.01.2021 Design according to construction manual	6	Measurement of conductivity during operation of the prototype	3
[higher-level systems] Finger pressure does not ring evenly	7	[Bell switch] Input and output line not completely electrically connected due to actuating pressure		Revision state: 15.01.2021 The occurrence could be reduced after measurement on the prototype and design review.	2		3
[higher-level systems] unwanted noises	6						

7.1.4 Answer own example

8 Appendix

8.1 B evaluation catalogue

Current abbreviated B rating table according to AIAG/VDA2019

10	Safety or existence relevant, user hazard
9	Relevant for approval (machine, customer, product)
8	Loss of a main product function
7	Restriction of a main product function
6	Loss of a product convenience function
5	Restriction of a product comfort function
4	Visual defects/increased NVH values - perceived by most customers/users
3	Visual defects/increased NVH values - perceived by many customers/users
2	Visual defects/increased NVH values - perceived by some customers/users
1	No perceptible effect

Complete B rating table according to AIAG/VDA2019

B	Impact	Criteria for significance
10	Very high	Impact on the **safe operation of** the vehicle and/or other vehicles, the **health of** the driver or passengers or other road users.
9		Non-compliance with **legal or regulatory** requirements
8	High	**Loss of a major function necessary** for normal vehicle operation over the intended service life.
7		**Restriction of a main function necessary** for normal vehicle operation over the intended service life **necessary main function**
6	Medium	**Loss of a comfort function**
5		**Restriction of a comfort function**
4		**Clearly perceptible impairment of quality of** appearance, sound, vibration, roughness or haptics
3	Low	**Moderately perceptible quality impairment of** appearance, sound, vibration, roughness or haptics
2		**Slightly perceptible quality impairment of** appearance, sound, vibration, roughness or haptics
1	Very low	No perceptible effect

8.2 A-rating catalogue

Current abbreviated A rating table according to AIAG/VDA2019

10	New technology, no standards, no best practices
9	New design, first-time application of standards, no experience
8	New design, few standards, not directly applicable best practices
7	New design, standards and design specifications describe basic design
6	Similar design, standards and design guidelines exist but are insufficient
5	Similar design, experience from predecessor, procedure evaluated but not proven
4	Almost identical design, best practices and standards
3	Detail changes in known design, best practices and standards
2	Almost identical / robust design, field experience, best practices and standards
1	Identical design, same application, design flawlessness clearly demonstrated

Complete A-rating table according to AIAG/VDA2019

A	**Occurring error cause**	**Criteria for occurrence**	
10	Extremely high	**New technology** without experience or under uncontrolled conditions.	No experience for verification / validation. Standards are not available. Avoidance measures cannot predict / are not available for performance in use.
9	Very high	**First application of** a design with **technical innovations**. New application/materials or changed conditions. **No experience** for **verification / validation.**	No targeted avoidance measures for the identification of performance under certain Anf.
8			Avoidance measures are not a reliable indicator of operational performance. Some standards and best practices exist that do not directly apply to design.
7	High	**New design based on similar technologies**. New application/ materials or changed conditions. **No experience** for **verification / validation**.	Avoidance measures provide limited information on performance. Standards, best practices and design rules apply to the underlying design but not to the innovations.
6		Similar **design** to previous using **existing technologies/materials.** Similar application with changed conditions. **Test or application experience available.**	Avoidance measures have a limited ability to avoid a cause of failure. Standards and design rules exist, but are insufficient to prevent the cause of the fault from occurring.
5	Medium	**Minor changes to previous design**, use of **proven technologies/materials**. Similar application, or conditions. **Test or application experience available.**	Preventive measures can reveal defects in the product in relation to the cause of the defect and give limited indications of performance.
4		**Almost identical design** with short experience of use. Similar application or conditions. **Test or field experience available**.	Preventive measures can reveal product defects related to the cause of the defect and probably show design conformity. Design adaptations to best practices, standards and specifications.
3	Low	**Minor changes to the known design.** Same application with minor changes to conditions. **Test or application experience availablewith** successfully completed test procedure **available**.	Preventive measures can detect product defects related to the cause of the defect and predict design conformity. Design complies with standards and best practices from previous designs.
2	Very low	**Almost identical mature design with long experience of use**. Same application with comparable conditions. Test or application experience under comparable conditions.	Preventive measures can reveal defects in the product in relation to the cause of the defect and demonstrate reliability of design conformity. Design is proven to meet standards and best practices from previous designs.
1	Extremely low	Cause of error is excluded by the design.	Error is eliminated through avoidance measures.

8.3 E-rating catalogue

Current abbreviated E rating table according to AIAG/VDA2019

10	No tests or test procedures (or too late)	
9	Test procedure not designed for specific consequence, fault or cause.	OK-NOK, Test-to-Fail, Degradation Test
8	New test method, not proven (E uncertain)	OK-NOK, Test-to-Fail, Degradation Test
7	New test method, not proven (E very low)	OK-NOK test
6	Detection of consequence, fault or cause proven but low	Test-to-Failure
5	Detection of consequence, fault or cause proven but moderate	Degradation test
4	Detection of consequence, fault or cause proven but moderately high	OK-NOK test
3	Detection of consequence, fault or cause proven and high	Test-to-Failure
2	Detection of consequence, fault or cause proven and very high	Degradation test
1	100% detection of fault or cause	Demonstrable

Complete E-rating table according to AIAG/VDA2019

E	Discoverability	Maturity of the detection method	Discovery opportunity
10	Very low	**Test procedure** has **not yet** been **developed**	No test method defined
9		**Test method not specifically designed** to detect the type or cause of the fault	OC-NOK, Test-to-failure, degradation test
8	Low	**New test method, not proven**	OC-NOK, Test-to-failure, degradation test
7		**Proven test method** for verifying functionality or validating performance, quality, reliability and durability; **planned use later in** the development cycle so that negative tests can lead to **production delays** due to design or tool revisions	OK-NOK
6	Medium		Test-to-Failure
5			Degradation test
4	High	**Proven test method** for verification of functionality or validation of performance, quality, reliability and durability; **planned use is sufficient** for adaptation of production tools for release for production	OK-NOK
3			Test-to-Failure
2			Degradation test
1	Very high	**Previous checks confirm** that the type or cause of the fault cannot occur or Detection methods **demonstrably** always **discover** the type or cause of the fault.	

8.4 AP Matrix

Currently valid AP matrix according to AIAG/VDA (J. Pfeufer / Pocket Power 2021) for design and process FMEA

		Discovery		
Meaning	**Appearance**	1-4	5-6	7-10
1	2-10	L	L	L
2-10	1	L	L	L
2-4	2-3	L	L	M
	4-5	L	M	H
	6-7	M	H	H
	8-10	H	H	H
5-8	2-3	L	M	M
	4-5	M	H	H
	6-7	M	H	H
	8-10	H	H	H
9	2-10	H	H	H
10	2-3	L	M	H
	4-5	M	H	H
	6-10	H	H	H

8.5 Form

The new form with all columns is confusingly large. Therefore, we recommend either the compact form (Excel) or the simple form (functions and system element in the context of the error, consequence and cause).

FMEA plus AKADEMIE	short Form Design-FMEA (requires context of the structure and the function in Effect, Failure and Cause levels)		
Planning and preparation (Step 1)			
Company:	Topic:	Side	of
Development site:	DFMEA Start date	DFMEA-ID:	
Customer:	DFMEA Revision date	Design Responsibility:	
Model/year/programm:	Interdisziplinary team:	Confidentiality level:	

Failure analysis with Structureelements and Functions (Step 2, 3, 4)				**Risk analysis (Step 5) and Optimization (Step 6)**						
1. Failure Effects (FE) next higher level	**Severity (S)**	**2. Failure Mode (FM) of focus elements**	**3. Failure Cause (FC) next lower level**	**Preventive actions**	**Occurrence (O)**	**Detection actions**	**Detection (D)**	**Action Priority (AP)**	**Status**	**Remarks**
						actions				

Official abbreviated VDA form:

FMEA plus AKADEMIE	Design-FMEA		Planning and preparation (Step 1)		
			Company:	Topic:	Side of
Structure analysis (Step 2)			Development site:	DFMEA Start date	DFMEA-ID:
1. Next higher level Ebene	2. Focus level	3. Next lower level	Customer:	DFMEA Revision date	Design Responsibility:
Function analysis (Step 3)					
1. Function higher level	2. Function focus	3.Function lower level	Model/year/programm:	Interdisziplinary team:	Confidentiality level:

Failure analysis (Step 4)				**Risk analysis (Step 5) and Optimization (Step 6)**											
1. Failure Effects (FE) next higher level	Severity (S)	**2. Failure Mode (FM) of focus elements**	**3. Failure Cause (FC) next lower level**	Preventive actions	Occurrence (O)	Detection actions	Detection (D)	Action Priority (AP)	Filtercode (Opt)	Name Responsible	Planned completion date	Status	Measure taken	Completion date	Remark
				Current actions											

8.6 Software

The market for FMEA software is relatively large. To find the right software for you, you should take your time and seek independent advice.
Criteria (examples) you can think about (no claim to completeness):

Must criteria:

- 7 steps: preparation, structure, function, measures, error analysis, optimisation and presentation can be carried out clearly, correctly and completely.
- AP: Evaluation according to Action Priority according to VDA2019
- Requirement, feature, function, process feature Distinction between requirements, functions and features: Consistency of features between system, design and process FMEA. Correct implementation of product / process features.
- Special handling features: consistency, transparency and handling.
- Evaluation catalogues: All standard catalogues (VDA, AIAG, etc.) available. Can be individually created / copied, sorting/grouping function Bonus for extra column for product-specific examples.
- BxA matrix (sequence level / e.g. medicine): BxA matrix, at sequence level (sequences have meaning and occurrence). Representation in tables, lists and matrix.
- Difference analysis (RPN, RMR, ...): Clear representation. Optimisable and differentiable.
- FMEA XML MSR interface: Easy and correct import and export via XML MSR format.
- Form VDA/AIAG/individual: All forms (VDA, AIAG, ...) included. Additional individual forms can be created (additional field contents can be determined via macros?).
- Graphical error analysis: Can be displayed and edited in a horizontal tree diagram and a horizontal hierarchy tree. Clear visual appearance and quick editing.
- Graphical function analysis: Can be displayed and edited in a horizontal tree diagram and a horizontal hierarchy tree. Clear visual appearance and quick editing.
- Graphical structure analysis: Can be displayed and edited in a horizontal tree diagram and a horizontal hierarchy tree. Clear visual appearance and quick editing.
- Hierarchical display: The structure and other attributes can be displayed in a logically structured and clearly presented hierarchy.
- Measure management: Developed measures must be consistently communicated and monitored with regard to implementation, effectiveness and verifiability. Task communication to project management. Measure tracking system and convenient import of processed measures. It should be possible to copy the same measures with the same status (1-click).
- Pareto analysis: Presentation of the highest risks (BxA, BxE, AxE, RPZ, RMR) according to the Pareto principle (80/20). Quick, clear and simple creation. Optimisable and differentiable.
- Manual logging option: functionality, clarity. Simple generation of dates, tasks and persons in the programme.
- Risk matrices (cause level): Correct, fast graphical implementation. Matrices can be optimised and differentiated.
- Cross-structure linking: Clarity and usability of linking several FMEA structures. (e.g. process and product FMEA).
- General team definition: Quick and easy definition of the FMEA team FMEA-related, direct transfer to the minutes, cover sheet and form. Definition of meeting team.
- Variant handling: Possibility of creating variants of previous FMEA and handling them correctly and avoiding errors as much as possible.
- Versioning: Creation of versions of the FMEA after editing with labelling (date, editor and scope of change, locking of the edition).

Shall criteria:

- 3D traffic light factor / RMR Correct determination from BxA, BxE and AxE (also RMR). Clear representation. Individually editable templates.
- Operation during moderation: Most important point. Software fast and easy to use, clearly laid out.
- Drag and drop modelling: Fast working in both representation forms by "drag and drop" so that complete strands can be repositioned or copied, multiple selection.
- Excel interface: Easy output of various FMEA analysis data in Excel format. (form, measures, individual, ...). Simple import and export of Excel files. (forms, measures, ...)
- Focusing and information reduction in graphics: Quickly focus on individual functions to support detailed discussions about a function. Hide and show additional information.
- Global measures: Definition, editing and overview of global measures at a central or arbitrary location.
- GRA (operating states): Different operating states of the system under consideration can be mapped, simple clear risk analysis possible at the beginning of the project.
- Groups of measures: Differentiation of measures into e.g. development, service and customer operations or different optimisation proposals for decision. It is evaluated whether and how these groups can be created.
- Offline editing: For "itinerant moderators" and home office.
- Automatic protocol: Creation of an automatic protocol of all actions performed (additions, changes, contents, etc.).
- Simultaneous processing: Correctness, operability, manageability, clarity in consolidation, behaviour in case of conflicts.
- Language handling: number of languages, scope and effort, translation options.
- Symbolic persons and appointments: Use of symbolic persons (project manager, designer, test engineer, etc.) and symbolic appointments (DP, PC, SOP, etc.).
- Merge / split: Measures (also errors and functions) with editing possibility, i.e. possibility to split / possibility to merge.
- Zooming: Fast zooming in or out of content to support discussions during moderation and presentation (also during video conferences).

Can criteria:

- Requirements management: Dealing and communicating with requirements. M to n links possible?
- Block diagram handling: Block diagram for efficient communication. Minimum requirement: attach to the FMEA or better take over or link elements directly (e.g. stamp). Connections recognisable. Representation of signal paths. Division of modules.
- Block diagram interfaces: Link to tools for block diagrams / system architectures (e.g. Enterprise Architect).
- CAD interface: Drawing can be imported directly from the CAD system or database, features can be transferred to the FMEA by stamping.
- Decomposition ASIL: ASIL: Integration of ASIL decompositions, also graphically visible.
- DOORS interface: Links possible? Usability, import-export possibilities.
- Error autom. linking: Error linking possible on the basis of the (existing) function link including selection options.
- Window technology, multiple monitors: Clear and individual presentation. Hierarchical representation, function tree etc. Freedom of arrangement of elements. Support for moderation. Second monitor. Fast arbitrary window arrangement.
- FMEDA: Input and evaluation of quantitative failure rates, failure types and diagnostic capabilities. Certification of the software according to the required standards (DIN EN 61508, ISO 26262). Link to the FMEA.

- FTA (qualitative/quantitative): miscalculations are excluded and methodological correctness is ensured for all levels of complexity and safety requirements. Generation with subsequent editing from error networks.
- html (batch): Output of all forms, all evaluations and results of the FMEA in HTML format. Possibility to save the batch order for intranet application.
- Integration into other management software: Integration capability in software (ERP, CAQ, FuSi, databases, document management systems, etc.)
- Ishikawa diagram: clear graph. Optimisable and differentiable. Hide and show additional information. Synchronised with all other views.
- Cost of measures: Input of estimated costs (comment, attribute) and evaluation to support prioritisation of measures and support decision makers.
- Costs Error, risk costs: Input of estimated risk costs (comment, attribute) and evaluation to support prioritisation of measures and support decision makers.
- Mechatronic modelling: Correct and practicable handling of operating situations, fault detections, fault reactions and fault consequences.
- Medical technology ISO 14971: Fulfilment of all requirements from ISO 14971 for risk analysis
- Batch processing to result in pdf files: Possibility of saving a freely definable batch job in a PDF file. Output of all forms, all evaluations and results of the FMEA in PDF format.
- Process flow chart handling: Process flow chart for efficient communication. Minimum requirement: attach to the FMEA or take over or link elements directly. Interrelationships recognisable. Representation of workflows. Division of modules.
- Test and control plan: Creation of test and control plan from FMEA (or vice versa). Common format of output (e.g. product characteristic, process characteristic).
- Useful shortcuts for editing: Software should also be operated quickly by the moderator via the keyboard. Operating aids and short cuts.
- Linking possibilities from outside: Targeted entry coordinates (e.g. on error, cause or measure from e.g. document or requirements management).
- Other interfaces: Interfaces with other external or internal programmes? For info!
- Word interface: Results from the analyses can be exported in Word format for further processing.

The most important things in a nutshell:

- FMEA preparation according to the state of the art (e.g. 7 steps, AP) is possible.
- Clarity for the team.
- Speed at which the FMEA facilitator can enter and model the statements.

In order to give you a small overview here, we have created a small compilation of the analysis software with which we have already worked and made experiences:

Focus on FMEA:

- APIS
- Audimus
- Plato
- QMH

Focus on safety:

- ENCO-SOX
- Ansys Medini Analysis

Focus on CAQ:

- AHP
- babtec
- Boehme and Weihs
- CAQ-AG
- CAT
- iqs
- Knowllence
- Pickert and Partner
- QMS (Siemens)

Other:

- SAP

And many more - although many are table-based and unclear and unsuitable for more complex FMEAs:

- AquaPro
- Datalyzer
- EnTeSo
- isograph
- MADe
- Matrix FMEA
- MBFG CIMOS
- med
- Orcanos
- pdap
- PQ-FMEA
- ProSeS (Quipsy)
- PTC Windchill
- QDA Solutions
- solvtec
- Syncos/Oxaion
- Synopsys
- Synostics
- t-method
- tqm

Final notes